John Beale Davidge, William Pechin

A Treatise on the Autumnal Endemial Epidemick of Tropical Climates

Vulgarly Called the Yellow Fever

John Beale Davidge, William Pechin

A Treatise on the Autumnal Endemial Epidemick of Tropical Climates
Vulgarly Called the Yellow Fever

ISBN/EAN: 9783337375553

Printed in Europe, USA, Canada, Australia, Japan

Cover: Foto ©berggeist007 / pixelio.de

More available books at **www.hansebooks.com**

A
TREATISE

ON THE

Autumnal Endemial Epidemick

OF

TROPICAL ▮▮▮▮▮**S**,

VULGAR▮

YELLO▮ ▮▮▮▮;

CON▮

ORIGIN, HISTOR▮ ▮▮▮CURE;

TO▮

A Few ▮

PROXIMATE ▮

‹‹‹‹‹‹‹

JOHN B. ▮

A. M.   M. D.

‹‹‹‹‹‹‹◆◈◆◈◆›››››››

ALTERAM PARTEM AUDI.

BALTIMORE:

Printed by W. PECHIN, *No.* 15, *Baltimore-ſtreet.*

1798.

Introduction.

IT is not my intention in the following pages, to
unfurl the banners of perfonal oppofition. Dr.
Rufh profeffes truth to be the object of his inqui-
ries—truth is the end of my refearches. The ex-
panded mind of Dr. Rufh, receives more fincere
and real pleafure in the perception of one truth,
than in all the fulfome incenfe that could afcend
from a thoufand flatteries : His great mind, I
hope, is clofed to the poifonous nutriment of boy-
ifh vanity.

The object of my labours and ftudies, for feveral years paft, has been, in a great meafure, an inquiry into the nature and etiology of difeafes.
And although, not unfrequently, I could meet
with the opinion, that contagious difeafes did often
originate in marfh exhalations, but as this opinion
was accompanied by feeblenefs, and want of conviction, I treated it more as the fondnefs for no-
velty and innovation, than as the candid refult of
experience. Nor fhould I at prefent attempt to
arreft its growth, had it not found too able an
abettor in the learned and elegant Rufh. Writers
of little note cannot give tone to opinions. Fal-
cons do not feed on flies. Great names alone can
introduce great errors.

Who among us would have imagined, that the
human mind was nothing but a flux of ideas, had

not David Hume given currency to this fancy? Or that thought was a peculiar configuration of the combination of material atoms, had not Dr. Priestly introduced it to our acquaintance? But, to come nearer home, who could suppose that the water spout, that huge rising of the ocean, was effected by a power of suction resident in the clouds, without the hint of Dr. Franklin.*

There has been of late, a great objection, to nosologies, started. I see no injury that they can operate : They unquestionably facilitate the acquisition of every science, as well that of Nature, as of Botany and Medicine. Nosology is a mirror reflecting, in pathognomonicks, the melancholy catalogue of cruel diseases ; and a student will learn more from a well arranged synoptical nosology, in one year, than he will otherwise in five. The labour of wading through bulky folios, and turning over musty aphorisms, is much more fatiguing and unproductive, than the attentive study of those so much asperfed *multa in parvo*. A *tyro* will much sooner acquire the diagnosticks, when arranged by the learned and experienced, than when he has to cull them out from a cloud of similars.

It is true that we should prescribe for the nature of symptoms ; and it is equally true that in prescribing for a symptom, we should have an eye to the nature of the epidemick that may prevail at

* The true and real cause of the water-spout is ; a column of electric fluid descends from the cloud, this electric column rarefies the air to a given distance, the rarefaction of the air causes an unequal pressure on the water : The equal pressure being removed, the water immediately rises in bulk corresponding to the space of the rarefied atmosphere. This column of water ascends in proportion to the levity or thinness of the air, sometimes to an amazing heighth ; and in its ascension acquires a whirl occasioned by the circumambient air rushing to a focus to fill up the partial vacuum and restore to itself its former equilibrium.

the time. The inference is, that, in prescribing for a nausea and puking in concussion and apoplexy, we are not to look at the insulated symptoms, but to inquire into the nature of the diseases ; in sickness from concussion no physician will think of the lancet, when he knows it to arise from a sudden derangement of the nervous power ; but for the same symptom in apoplexy he immediately lets blood, from a knowledge that it proceeds from compression of the brain.

We must always practice according to the nature of symptoms, which nature is only known from a knowledge of the nature of the disease. For a dark tongue in a typhus we order good wine, for a dark tongue in the yellow fever we order mercurial purges : The symptom is the same, but experience informs us that the same remedies will not answer. Then let physicians be cautious how they prescribe blood-letting for every giddiness— the giddiness in the jail fever is cured by wine, in the yellow fever by active purges and the lancet.

Great names have a species of magick accompanying them. Mr. John Hunter gave shape, for a time in London, to the doctrine of digestion ; his opinion, that it is the combined action of friction and fermentation, has been successfully combated by being put in competition with one better founded. Spallanzani, by a series of well managed experiments, has exhausted the subject of all doubt ; according to this indefatigable man, it is nothing more complex than a simple chymical process ; such as takes place in the union of sugar and water, where the attraction of combination overcomes the affinity of aggregation : That which is proper nourishment is taken up, by the lacteals acting as capillary tubes, what remains is thrown off as excrementitious refuse.

Nature has placed limits between diseases, as

weil as between vegetables, and animals, nor muſt we ſuffer the charms of ſimplicity to perſuade us to intrench on thoſe outlines. The ſmall-pox is generically different from the varicella. By taking away natural boundaries, we ſhall bring the ſciences, in confuſion, about our ears, and precipitate ourſelves into a chaos of uncertainties. A diſreſpect to civil diſtinctions, brings about political demiſe.

Animal electricity, in its haſty progreſs, promiſed us an acquaintance even with animal life and human phyſiology; and had we all been Italians, we might, before this, have found the philoſopher's ſtone: The next thing would have been, the good luck of Paracelſus, an introduction to an univerſal panacea.

This comes forth, neither connected with name, nor clothed with title; I leave it to recommend itſelf. If it proſpers, well; if not, it will only, ſuffering the miſhap which overtook one of David Hume's beſt productions, fall dead-born from the preſs.

Names frequently are more propitious to books than their contents are. And many works, of ſterling value, too often inſtruct none but their authors.†

† When the above was written I did not intend to have put my name to this production, but for particular reaſons I have changed my firſt intention, as the reader will perceive.

THE
CONTENTS.

CHAPTER I.

The Origin, History, &c.

CHAP. II.

Proximate Cause analyzed.

CHAP. III.

Diagnosis.

CHAP. IV.

Cure.

TREATISE, &c.

CHAPTER I.

THE ORIGIN, HISTORY, AND NATURE OF THE YELLOW FEVER.

THE yellow fever, fynonimous with la maladie de fiam, or la fievre matelotte of the French, and vomito prieto of the Spaniards, is the majority or acmé, as the intermittent is the embryo, of the remittent bilious fever ; It is to the common bilious, what the confluent is to the common mild fmallpox ; they are in kind the fame, a fpecific difference only exifts between them. It is conceived in the fame matrix, and quickened by the fame fun.* It is indigenous to America and all other warm climates. It is the great outlet, to Americans and Britons, from life to the grave.

The rays of the fun diffufed and fcattered are falutary and innocent, but collected and condenfed into a focus are dangerous and hurtful : thus the concentrated effluvia of marfhes is deadly and venemous ; but fcattered and diffufed, circulate among us innocently.

The generality of the French writers call it la maladie de Siam from a fhapelefs notion that it was originally from Siam, a country in the eaft : this contains as much truth as the opinion that it attacks failors only ; whence they call it la fievre matelotte. It is, in all probability, the caufus or

---

* " Without the matrix of putrid vegetable matters, there can no more be a bilious or yellow fever generated amongft us, than there can be vegetation without earth, water or air." Rufh vol. 3. p. 168.

febris ardens of Hippocrates, Aretœus and Galen. Trallian and Lommius appear to have feen this fever.

Ulloa makes mention of the vomito prieto prevailing, in a moft horrible and deftructive form, in Carthagena in the year 1729 and 1730*. It made its inroads in Barbadoes in the year 1696† a time long prior to the vifit of Dr. Warren to that ifland.

Marfh effluvia appears to be either the decompofition of vegetables or water, whether infcrutably combined with fomething elfe, or infulated, I leave for farther inveftigation : in the decompofition of either, hydrogene is produced in confiderable quantity. Water, in its liquid ftate, is a compound of hydrogene and oxigene, with an addition of caloric and a little common air : Its decompofition is effected by the rays of heat inferting themfelves between the liquid particles until their feparation is fuch as to annihilate the attraction of combination between the hydrogene and oxigene. Whether then an union, betwixt the oxigene and light, comes to pafs, remains to be difcovered ; this I ftrongly fufpect. The hydrogene being difengaged and infulated, and very much accumulated, or peculiarly combined, perhaps becomes the peccant agent.

It is worthy of obfervation that the bilious or yellow fever does not generally prevail during the heat of the fummer ; this may be owing to the greatnefs of the heat diffipating and fcattering the hydrogene, or marfh effluvia, fo as to enfeeble and render it innocent.

That a combination between the oxigene and light happens, is likely, firft from their natural af-

* Voyage to South America, B. 1, C. 5.

† Hughs' Hiftory.

finity to each other, and fecondly from a phœnomenon obfervable during ignition ; the rays of
light falling immediately upon the ignited com-
buftibles, caufe the flame to become faint and ul-
timately will extinguifh every particle of the fire.
The probability is, that this phœnomenon is occa-
fioned by the rays of light attracting and combining
with the oxigene of the atmofphere, and thereby
interrupting the procefs which was going forward
between the oxigene and combuftible bodies. To
this there is an acceffion of additional ftrength
from what takes place in vegetation.

A vegetable, kept in the fhade, becomes white
and fickly ; when it is expofed to the light it re-
vives and becomes healthy : this I apprehend flows
from the light acting as a ftimulus, and at the fame
time attracting from it its oxigene, with which it
is neceffarily charged in decompounding water for
its nutrition. Vegetables when analyzed yield
more or lefs of hydrogene. Hydrogene is that gas
which in its ftruggle to afcend, meets with the elec-
ric fluid of the atmofphere and forms what, in ver-
nacular vulgarifm, is called jack-o-lanthorn—or
when it has gained the fuperior regions and formed
the upper ftrata of the air, comes into contact with
the electric fluid, and effects what enjoys the appellation of aurora borealis or northern lights. It
is alfo the principal agent in the motion of the
aeroftatic machines.

The yellow fever made its firft appearance in the
city of Baltimore in the laft of Auguft. The common bilious fever prevailed at the Point from June.
A lady from Philadelphia, bringing with her the
feeds of the difeafe, which were brought into action by the fatigue of the journey, was feverely
attacked with it in Charles-ftreet, fhe had the ge-
nuine black vomiting, which refembled ink and
coffee-grounds mixed, for two nights and a day,

miscarried on the sixth night of the disease—she notwithstanding recovered ; no person in the family, nor neighbourhood had it during the whole season.

This, together with the number of cases of violent bilious fever at Fell's Point, threw the city, generally, into a combustion ; the committee of health requested that the physicians would convene in order to give a report of the city ; upon the convention of the physicians, it appeared, from their joint testimony, that the above-mentioned lady was the only person. labouring under the fever in the west end of the city, called the town in contradistinction to the Point. [I shall in what follows make use of the distinction of town and Point] The committee received a letter from doctor Coulter, a physician of great respectability at the Point. Doctor Allender, in person, waited on the committee, and gave information of that part of the city. It was requested by the committee of health, that some of the physicians of the the town would visit the Point ; pursuantly to this request, doctors Goodwin, Moores and Davidge politely went, and waited on the sick individually ; they were piloted by a gentleman who lives with doctor Allender, together with a student of doctor Coulter's. Their report was, " that there was nothing more than a violent remittent bilious fever prevailing :" This perfectly accorded with the sentiment which dropt from one of those physicians during his report to the mayor, " his reason (he said) why he did not report the case of the lady of Philadelphia, was a full persuasion, that the yellow fever, being of the same origin with a violent bilious fever, could not be multiplied, by an intercourse of bodies, under any circumstances whatever."

The above report is a strong and prominent feature of their discernment, and indifference to popu-

lar prejudices. A little after that period, the difeafe rifing from the grade of a bilious to that of the
yellow fever, mounted its chariot of death, and
drove furioufly through the ftreets, fowing difmay
and mourning wherever it approached : conveyed
by the north-eaft wind it fcattered itfelf all along
Federal-hill, and weft end of the bafon ; which-
ever direction the miafmata (arifing from the ftagnant water and marfhes about the Point and
wharves) controuled by the winds, took, the difeafe
tread clofely in its footfteps. It evolved, in an
horrid and difmal fhape, its venemous difpofitions
in the fouth end of Hanover-ftreet and its vicinity.
After a fhort interval this deadly effluvia penetrated
into the vitals of the city, and many who were not
near the Point nor wharves, thofe exuberant fountains of mifchief, took it, either in their houfes, or
in the ftreets. Not an inconfiderable number of
thofe who were at the launching of the frigate received the difeafe, feveral of whom died of it afterwards up in the town ; but, fortunately for the
citizens, no perfon was infected from them. A
very confiderable part of the inhabitants of the
Point fled into the country, and fome from the
town removed ; a temporary defert was effectuated.

I fçarcely need mention the daily deaths ; the
reports of the committee of health are inadmiffible
of an acceffion of teftimony to give weight to their
authenticity and accuracy.

The greateft number, in the the the town or
weft end of the city, in any one day was feven or
eight ; the lift of the deaths at the Point was, for
fome time, confiderable. This fever began to
faint about the firft of October, and was nearly or
quite extinct on the firft of November.

All endemical epidemicks, as they depend on a
peculiar conftitution of the air, muft of neceffity

ceafe whenever this condition is deftroyed, whether it be by frofts or rains.

The yellow fever, like all other epidemicks, delights in folitude. An epidemick, whether endemical or contagious, depends on a general peculiarity of the air ; which general peculiarity will have a general influence within its own dominion, and communicate a general aptitude to all bodies, within this jurisdiction, to receive its action. It chafes away all others of leffer ftrength, as is juftly obferved, firft by Diemerbrook, fecendly by Sydenham, thirdly by Pouppé Defportes, and after them by Rufh ; who is perfectly correct where he fays that no two epidemicks, of unequal force, can prevail at the fame time. How is it poffible for two general and oppofite conftitutions of the air to exift at the fame time? with equal propriety we would fay that a cord can enjoy two diftinct ofcillations at the fame inftant ; or that two particles of matter can occupy the fame given fpace in the fame divifion of time. But this every tyro fhould know.

Some of the firft traces of the yellow fever in America are to be found about fixty or feventy years back. A phyfician, in converfation, the other day told me that he had met with the yellow fever in Baltimore ever fince he had lived in it, which is fifteen or twenty years. It is violating all obligations of decency and truth to fay that it is of recent date. The town of Baltimore, in proportion to its inhabitats, is lefs fubject to this autumnal remittent or yellow fever, than the low fituations about the Potowmac. A gentleman, who for fome confiderable time was one of the principal directors of the Potowmac bufinefs, informed me that one feafon they loft a confiderable number of their workmen by the above fever. And that in no inftance did it fpread by contagion.

Difeafe is fimple, and indivifible ; it is the abfence of health. But, for the facility of the pen, and eafe of fpeech, we make difeafe an aggregate. It is not, at all times, proper to act the philofopher and metaphyfician; we muft accommodate ourfelves to popular ufage. Cold is the negation of heat, and darknefs the abfence of light. Ice is the natural ftate of water : yet cuftom, the ftandard of all languages, perfuades us, in familiar converfation and common writing, to fay it is cold, or it is dark, &c.

Every country has difeafes proper to its climate and fituation. Some difeafes are common to every country and climate : Accident and particular circumftances will create fporadick difeafes, in every country, not peculiar to them refpectively ; a difeafe proper to one country can, by the medium of intercourfe, be carried into another and there propagated.

Britain has its fcrofula, and typhus ; the vicinage of the Alps has its goiter : the Eaft has its plague ; the Weft-Indies, America and other countries within or near the tropicks, have their remittent bilious fevers, and hepatick affections. It is not the import of this paragraph, that thofe difeafes are exclufively generated in thefe countries refpectively. The antogonift idea is incontrovertable. But as thofe complaints, although they may originate or be produced in every country and under every climate, moft commonly and generally appear in thofe countries in the manner above-mentioned, I have taken the liberty to ftile them proper to thofe individual climates. It is a concurrence of circumftences, and not given latitudes, that is requifite for the production of difeafes.

The feeds of the remittent bilious yellow fever are engendered in putrid vegetables and ftagnant water, and are vivified by heat and drynefs. Thofe

C

who live in their neighbourhood, in warm climates, and during hot and dry autumns generally suffer more or less. The more elevated situations, and those distant from such sources in the West-Indies and America, most generally escape this deathful malady, except when the atmosphere becomes universally surcharged with the effluvia emanating from those exuberant fountains of mischief and poison; under such occurrences they unavoidably participate of the evil. This is echoed from the united voices of writers and practitioners.

Males, by being more exposed to this effluvia floating in the air, to the violent rays of the sun, to night-dews, and to the various, sudden, and great vicissitudes of the weather (these operate differently according to their respective natures) are more subject to this, and all endemical epidemicks, than females, whose business is naturally within their houses, where the poison is blunted and rendered inert by the fires and smoke, and has itself dissipated by striking against buildings.

The skirts of most cities are occupied by the poorer class of inhabitants; their houses are exposed to the first and most violent assaults of all endemical epidemicks; but as those epidemicks intrench more the vitals and heart of a city, they are dissipated, enfeebled, and disarmed.

The remittent bilious yellow fever neither pities helpless infancy, nor reverences the decrepitude of age; but its chief delight is to jostle, from the stage of life, the young and vigorous   The many headed monster, armed with destruction and woe, wantons in the citadel of life: and now, Anthropophagi like, without warning or prelude implants his merciless dagger; and now under the garb of innocence, gambols in the cheek of health.   The yellow fever is as fatal when with the investiture of an intermittent, as when habited in its regular type. Having feen its birth and extract we make an eafy ftep to its nature.

<table>
<tr><td>Dr. Ruſh.</td><td>The Author.</td></tr>
</table>

Dr. Ruſh.

"The fevers generated by putrid cabbage, mentioned by Dr. Rogers, and by putrid flax mentioned by Dr. Zimmerman, were both contagious. Dr. Lind afcribes the yellow fever every where to marſh or vegetable exhalations; and this fever, we know fpreads by contagion. Dr. Lind jun. eſtabliſhes the contagious nature of the marſh fever which prevailed in Bengal in the year 1762. I ſhall tranfcribe his words upon this fubjeċt. " Although marſh miafmata (fays he) firſt bring on the difeafe, yet contagion prefently fpreads it, and renders it more epidemic. Thus the Drake Indiaman continued free from the diforder for two weeks together, when ſhe had no communication with the other ſhips; whereas as foon as the diforder was brought on board, many were feized with it within a few days in fuch a manner as to leave no room to entertain the leaſt doubt concerning its peſtilential nature,"

Dr. Clark mentions a contagious malignant fever from marſh miafmata. which prevailed at Prince's Iſland in the year 1771, and which afterwards infected the Grenville Indiaman. The contagious peſtilential fever in France, fo accurately defcribed by Reverius, was produced by an exhalation from putrid vegetables, particularly hemp and flax. Even intermittents, the moſt

The Author.

An endemick is a difeafe, that afflicts feveral people together in the fame country where it reigns, ariſing from local circumſtances or a peculiar condition of the air. It cannot be carried from one country to another, by the means of bodies and cloths. It affects more or lefs all within its own periphery. It cannot be carried, beyond its own atmofphere, by bodies difeafed with it. No difeafe ariſing from marſh effluvia can be communicated beyond the atmofphere charged with this effluvia. Every difcafe ariſing from marſh effluvia, from the gentle intermittent to the furious yellow fever inclufive, then is endemick.

It does not appear, that any of the cafes, in the antagonizing column, communicated themfelves by means of the difeafed bodies beyond the circumference of the atmofphere impregnated by the vegetable exhalations; then if they did not, which is pretty obvious, act beyond the limits of the inquinated atmofphere, the probability is, that the vegetable or marſh effluvia wafted through the air did the mifchief, and not the intercourfe of bodies.

That a ſhip's crew was free from a difeafe this week, is no juſt argument, if they be infected next, that it muſt be by means of one difeafed body communicating it to another; the contaminated air may have enlarged its limits—a particu-

frequent and moſt numerous offspring of the marſh exhalation, are contagious. Of this there are many proofs in practical authors. Bianchi deſcribes an intermittent which was highly contagious at Wolfenbuttle in the year 1666. Dr. Clark mentions a number of cafes in which this mild fpecies of fever was propagated by contagion.

Dr. Cleghorn has eſtabliſhed the contagious nature of intermittents by many facts. After mentioning numerous inſtances of their having ſpread in this way, he ſays " Theſe tertians have as good a right to be called contagious as the meaſles, ſmall-pox, or any other difeafe." &c. *Vol.* 3, *p.* 160.

lar direction of the wind may have conveyed the bad air to the ſhip. Its happening poſteriorly to a communication with the difeafed, is no argument, except at the fame time it be proved that this crew was without the controul of the ill-conditioned air. Marſh and vegetable effluvia ſmite at the diſtance of miles, this no phyſician doubts.

" Theſe difeafes (ſpeaking of intermittents) make their firſt appearance in February and Auguſt particularly; tho' fometimes they appear fooner or later, according as the air is more or lefs diſpofed to produce them, which, of courſe renders them more or lefs epidemic." P. 51, Dr. Sydenham.

A queſtion of importance and magnitude here emerges and invites our attention; it may ferve to awaken a little our judgments and elucidate the preſent ſtage of the buſineſs. Why is it that the intermittent fever has never clothed itſelf in its contagious habiliment in America. It is rather problematical that it has ever done more miſchief, or been more common and violent in any other country than in America. Here is a myſtery equally dark on all ſides; how are we to decipher the enigma?

" *Intermittentes.*

Febres, miafmate paludum ortæ, paroxyfmis pluribus, apyrexia, ſaltem remiſſione evidente interpofita, cum exacerbatione notabili, et plerumque cum horrore redeuntibus, conſtantes: paroxyfmo quovis die unico tantum." Dr. Cullen, Tom. 2, Synop.

Dr. Cullen, the luminary of the medical world, has been careful to inform us, that he has never met with contagious difeafes arifing from vegetable putrefaction.

Sydenham, ftiled by Dr. Rufh, " the incomparable phyfician," pofitively afferts that an intermittent becomes epidemick from the *air*, and not by contagion. Dr. Jackfon fays that the intermittents and remittents become epidemick and fpread by the marfh miafma being diffufed through the atmofphere, and not by contagion (vide Jackfon on the difeafes of America). Dr. Gilchrift mentions in his tracts on fea voyages, that the marfh effluvia being fcattered through the air becomes the fource of popular difeafes. That difeafes, flowing from marfh miafmata, are incapable of becoming contagious has been confidered, by the generality of phyficians, as a fact almoft felf-evident. And I believe that few will long hefitate to determine between the authority of Sydenham and Cleghorn.

The unfcientifick notion of intermittents and remittent bilious fevers being contagious, is too ridiculous to find accefs even to the eafy credulity of the unread peafant : but yet it is embraced by philofophick refinement. And what abfurdity, though ever fo enormous, has not at one time or another offended the dignified pride of delicate philofophy ? organized abfurdities, have long confpired againft the progrefs of fcience, and tyranized over the tender germes of jufter fyftems.

Endemick is the antithefis of contagious. Endemick and contagious eftablifh two oppofite categorics.

Contagion is the emiffion, from body to body, by which difeafes are communicated. A contagious difeafe arifes originally from human effluvia, can be carried, by means of the infected bodies, or cloths, from country to country, from city to city, from town to town, in fine from any one place to another ; The yellow fever has not its birth from

human effluvia, cannot be fpread, by means of difeafed bodies, or wearing apparel, or bed-cloaths,
from country to country, from city to city, from
town to town, nor from one place to another—the
yellow fever therefore is not contagious.

Either an endemick or contagious difeafe, becoming general, and affecting a whole country, or
great extent of territory, claims, in technical lan-
guage, the appellation of epidemick. It is not
whether a difeafe arifes from vegetable or human
effluvia, but whether it has a general or univerfal
action, that conftitutes an epidemick. It is the uni-
verfality of its action, and not the nature of the
fource whence it fprang, that ftamps its character;
whence two claffes of epidemicks—the endemial
epidemick, and the contagious epidemick. The
yellow fever ranges under the former, the plague
under the latter.

Doctor Rush.	*The Author.*
" It has been remarked that this fever did not fpread in the country, when carried there by perfons who were infected, and who afterwards died with it. This I conceive was occafioned, in part by the contagion being deprived of the aid of miafma-ta from the putrid matter which firft produced it in the city, and in part by its being diluted, and thereby being weakened by the pure air of the country. During four times in which it prevailed in Charlefton, in no one inftance, according to Dr. Lining, was it propagated in any other part of the ftate." Vol. 3, page 157.	When the intercourfe of bo-dies, labouring under a fever, cannot, without adventitious auxiliary, fupport and multiply the faid fever, the fever is not contagious; the yellow fever cannot fimply and without the addition of the effluvia arifing from vegetable putrefaction fupport and multiply itfelf, the yellow fever therefore is not contagious.

A little attention to the nature and operation of
the fmall-pox, meafles, or jail-fever, when conveyed into the pureft atmofphere, will, in fome
meafure, obviate the difficulty which obfcures the

above fact in relation to the yellow fever. It is found upon experiment, that, uniformly the small-pox, measles, or jail-fever require nothing more than their own presence and virulence to perpetuate themselves. This is luculently apparent from the many melancholy instances of whole families being precipitated to their graves by the unfortu-nate introduction of servants, purchased from on board ships, infected with the typhus or jail-fever. Out of fifty persons, who might visit a patient la-bouring under the latter stage of the small-pox or measles, forty-nine, in all probability will be in-fected. I have selected those diseases, as they are common and known to every body.

In these diseases, which are essentially and abso-lutely contagious, no heterogeneous aid is necessa-ry; in the yellow fever the addition of the marsh miasmata seems to be the *fine qua non*, in mul-tiplying and propagating the complaint, and the efficient, occasional cause of its rise, independently of any assistance from bodies : and without the help of this miasmata, the bodies cannot generate nor spread the disease ; hence it appears that the miasma produces the disease whenever and where-ver it may occur, and not the bodies labouring un-der the complaint.

It is not my expectation to turn, upon the axis of a syllogism, the whole esculapian world—preju-dice may for a time barr the understandings, but time and accident will eventually unlock the judg-ments of men.

Doctor Rush.	*The Author.*
" Let it not be inferred from the enumeration of the means of preventing the contagion of this fever, that I admit a con-tagious nature to be one of its characteristick marks. Far from it. It is an accidental circumstance produced chiefly	A disease which is not es-sentially, and in its own nature contagious, is not, in strict pro-priety, contagious at all—the yellow fever is not essentially and in its nature contagious— the yellow fever is therefore not contagious at all. A dis-

by the concurrence of the weather." And on the fame page. " It is in no inftance contagious in fome cafes." Vol. 4, page 61.

cafe which does not embrace and infold, in its very nature, radical, and inherent contagion, naturally repudiates the idea of contagion. Fire, elementarily, contains heat; and light is, intrinfically, luminous.

If the vegetable effluvia does and can, of itfelf, produce, perpetuate, and multiply the yellow fever : and if the human effluvia cannot and does not in any inftance, of itfelf, produce, perpetuate, and multiply this fever, upon their fortuitous union ; which is it that effects the mifchief? The utmoft precipitancy cannot endanger a decifion.

The bodies difeafed certainly produce an effluvia, which, by difturbing directly or indirectly the bowels, ftomach, or fenforium commune, may deftroy the equipoife of the animal fyftem, and thus prove an exciting caufe, in like manner with drunkennefs or night-expofure ; hence the aptitude of thofe who have to nurfe, and wait on the fick, to be difeafed. This is not by receiving it by way of contagion : but under thofe circumftances the equilibrium of the body is deftroyed, the œconomy is difarranged, the vigour is unnerved, and an advantage is afforded the miafma to bring itfelf, with all its lethiferous appendages, into exiftance.

The government of a country is called good or bad, according to its nature ; the beft government becomes hateful and deteftable from bad adminiftration ; but the adminiftration is not phyfically confecutive of the government. In this inftance we are to depofe our minifter, not change our civil fyftem. Thus if we remove the patient, in the yellow fever, from the inquinated atmofphere, he is no longer dangerous or hurtful to the attendants.

Chriftology, the fubftratum of hope and life, has become hoftile to the peace and fafety of nations by its adminiftration, not by its nature, its eflence is immortality and quiet.

Many things, in their nature, elude the moft vigorous effort of the human intellect, and tantalize the grafp of genius itfelf. They prefent to us their modes, qualities, and effects; thefe, operated on by the medium of the fenfes, advertife us of their refpective fubftrata. It is from the effects of human, and marfh effluvia, that we can have any clue to the fecret of their nature. Like caufes will always produce like effects, provided they operate upon the fame order of patients. The human effluvia eftablifhes one order of difeafes, the marfh effluvia another, thefe orders never embrace—they have no intercourfe. Of thefe identity forms the curve, and thofe effluvia the afymptotes, they apparently approximate, but they can never come into contact. The human effluvia can never produce the remittent bilious or yellow fever, nor can the marfh effluvia ever give origin to the typhus or jail fever.

Is it good logick, that a fever can arife from one fource to-day, and from another, diametrically and phyfically oppofite, to-morrow? This is certainly at variance with common ratiocination. Yet this muft come to pafs, if the yellow fever, originally, has its birth from vegetable putrefaction, and is afterwards perpetuated by contagion. Some have afferted that heat and cold, though different, produce the fame effects. Cold is a relative term—it has no abfolute exiftence—and that which has not an abfolute exiftence cannot poffefs a pofitive action.

Sir Ifaac Newton, who was a man of no humble genius, is of the opinion, that it is incompatible with the principle of philofophizing, in the expli-

D

cation of any phænomenon, to adduce more caufes than are abfolutely neceffary for its folution.

After the above theoretical endeavour, we fhall direct our attention, a little, to the authority of obfervation, and minute inquiry, of fome of the moft refpectable writers and practitioners.

It is an unqueftionable verity, a verity of the moft publick notoriety, that not one, of the very great numbers who have left the cities and towns, fome of whom have died and fome have recovered, has communicated the yellow fever to thofe who have attended. Not one folitary fact has ever reached me ; and my fcepticifm is fuch as to lead me into a perfuafion, that there has not exifted one unequivocal, well analyzed fact of a patient, going to the country, and there multiplying this fever.

" It affects the inhabitants of cities, and not of the country, as in Charlefton in the year 1732, 1739, 1745 and 1748. And in Philadelphia in the year 1793." That is, it could not be propagated, by the difeafed bodies, in the country. But that it can affect inhabitants of the country and alfo originate in the country is fully teneble. I know that my opinions here, as in other parts of this effay, are the antipodes of thofe generally received. But while I have facts, and thofe of the moft ftern and inflexible nature, I fhall not be over folicitous about received whims whether popular or medical.

On the Potowmac Bottoms, and along the Monokocy, I have feen the moft unequivocal cafes. It is alfo, from the very refpectable authority of Dr. Watkins, very common in different parts of Kentucky. Dr. Watkins was in Baltimore during the prevalence of the fever this autumn, and, with me, waited on the fick. He, after being provided of all neceffary requifites to form an opinion, declared it to be precifely fimilar to that of Kentucky. He farther obferved that the fever in that ftate frequently was attended by the black vomiting &c. He said that it was confidered there as an endemick arifing from the marfhes, and in no inftances contagious.

It is true that Dr. Rufh mentions an inftance of this fever fpreading in the country; but we want graver authority than inaugural thefes afford. Thefes, we all know, generally, are nothing but the echo of the preleƈtions of a preceptor.

In the flourifhing and growing city of Baltimore, we have had the moft ftubborn, and irrefragable proofs, of the yellow fever being incapable of fupporting itfelf, in the cafes which have occured about the wharves and Fell's Point; after thofe cafes were removed up into the city, they had their virulence to die with them, thofe who died; and from thofe, who recovered, all mifchief and fuppofed contagion evanefced into the empty air, which bore it to the pages of medical writers, not to the bodies of healthy attendants. This was the refult in 1794 and 1797.

The unfortunate cafe, of the very refpeƈtable Dr. E. Johnfon, with feveral cotemporary incidents, afforded a fhort-lived triumph to thofe who were wedded to the contagious fyftem; but when their opinions, armed with all the addrefs and fubtlety of the authors, came in collifion, with thofe of more ereƈt and manly afpeƈt, they felt their vacillating uncertainty, and ceded in candid conteft.

Which is moft confonant with probability, for gentlemen, going to an atmofphere, charged with poifon and death, an atmofphere immediately wafted from the Point and concentrated under a hill, where many had been forced, by its deathful influence, to pafs the bourn of life, and others to perceive death, with its black wings, to hover about them, to take it from thofe exanimous and moribund bodies or this death-bringing air? Taking into view, at the fame time, the fact, that none of thofe gentlemen, feveral of whom died up in the city, communicated, to any of their families or attendants, the difeafe.

Some even point out the lucklefs moment, in which the relentlefs malady feized them—at a particular juncture, I perceived fomething fingularly offenfive—in three hours after that inaufpicious point of time, I became unwel : another was taken ill a day or two pofteriorly to fome ill-omened hour, in which the breath of the patient was breathed directly into his face, it was then he inhaled, with the pabulum of life, the fermenting leaven of death. Who, after thefe and fuch like melancholy tales, can hefitate to believe the yellow fever contagious ?

I would no longer quarrel the fentiment of contagion, had one of the above occurrences taken place, in people, who were not. at the time of the imagined infection, in an impure atmofphere, or had not lately been expofed to air impregnated with the deftructive miafmata. The fact is, thofe perfons were all in this ill-conditioned air. I met with two cafes, in two young men who had been at the Point at the fame time, and were both attacked, on the fame day, with all the regular fymptoms of the fever, where the miafmata did not come into action until the eighteenth day—they were not, during the intermediate fpace, expofed to either fick bodies, or vegetable efluvia, having been in the country the whole time.

Whatever affects moft our fenfes, we are prone to attribute our evils to : This is excufable in men unaccuftomed to thought, but it is an incongruity in the common order of reafoning, and is at total variance with the notions of men of letters.

Logomachy is as inconfonant with my habits, as any two oppofites in nature are inconfiftent with each other; but I do not confider a war of fentiments fraught with mifchief or good, a colluétation of principles, tending either to fupport or deftroy the commerce of a nation, as the idle jar of words. It is no new aét for men to force the moft oppofite and contending principles into the moft cruel and unnatural union. Tradition, let it militate ever fo much with common fenfe; and an affociation of ideas, bearing no affinity nor cognation to each other, prefs us, tco frequently, into moft abfurd beliefs and habits of thinking. I might here caufe to pafs in review before the reader's mind the great Lavater's philofophy of faces, that finifhed and po- lifhed phyfiognomy of folly. And where is a phi- lofopher, bating a few, who does not ftuff his works with the infinite divifibility of matter? who has dared to quarrel the corollary, of the great Locke, that the human mind is, at firft, as a blank fheet of paper, paffive to the charaéters of chance? and where are the friends of David Hume, who do not, to this day, believe, that, if an afs were placed between two cocks of hay, which impreffed equally, he would ftarve, being unable to make a choice? It is a good hint, that we fhould tread lightly on the afhes of the dead; I feel its influence, and may David reft in peace, and his errors fleep in eternal filence.

Now were I to plunge into the vacuum of meta- phyficks, or enter the lifts in the gipfy-jargon of pneumatology; I fhould believe, with the peerlefs Reid of Glafgow, that the human mind poffeffed, inherently, action, vigour and choice; that it ope- rated upon furrounding objeéts, and was not the paffive fport of incidental impreffion. Whence comes genius, the innate perfpicacity of the hu- man intelleét; genius, that heavenly offspring, at all times impatient of the trammels of control, and indifferent to the habits of education? not from

the impreſſions of the beauties of Thompſon, nor
the more ſublime of Homer. It is in-born. Did
the mountains of Switzerland, give exertion to the
mind of a Haller, or the banks of the Potowmac,
infuſe divinity into a Waſhington? that encyclope-
dia of virtue and greatneſs, in whom may be found
every ornament of human excellence; into him
who firſt boldly dared, to rend aſunder the ſtrong
ligaments of prejudice, to control the imperious tide
of ancient uſage; who magnanimouſly deſpiſed
the boiſterous torrent of vulgar obloquy, and chal-
lenged the herald of recording time? Collateral
arguments are, frequently, as relevant as thoſe a-
riſing immediately from the ſubject.

This apparent digreſſion is to ſhew, that no name
is above truth, that our care and ſolicitude ſhould
be the inveſtigation and conſervation of truth, not
the ſupport and protection of names and traditions.

Doctor Ruſh.	*The Author.*
" An aptitude or prediſpo-ſition from ſeaſon, climate or conſtitution muſt concur to ren-der the contagion of this, as well as other malignant fevers, ſufficiently active to produce diſeaſe; as well might a travel-ler attempt to deſcribe the cli-mate of a new country, from the hiſtory of a ſingle ſeaſon, as a phyſician to fix the charac-ter of an epidemic from its ap-pearance in one ſeaſon, or one country." Vol. 4, page 62.	" The one (the yellow fever) is evidently cauſed by marſh effluvia heat, violent exerciſe in that heat, thick, hot, moiſt atmoſphere; night air and dews, and the abuſe of ſpiritous liquors. The other (the bou-lam fever) on the contrary, is cauſed by contagion alone. This is certainly the moſt re-markable difference; and con-ſtitutes an obvious, clear, and indiſputable diagnoſis." Chiſ-holm, page 147.

" But I never could obſerve any one inſtance,
where I could ſay that one perſon was infected by,
or received the fever from, another perſon who
had it." Hillary, page 145.

Dr. Jackſon calls the yellow fever of Jamaica
an endemick, and no where mentions its being
contagious; who alſo is of the opinion that moſt

epidemicks fpread by means of the marfh effluvia, and not by contagion.

Dr. Mofely ftiles it (the yellow fever) an endemial caufus, page 391. And ridicules the idea of its being malignant, peftilential, and contagious, as afferted by Dr. Warren, page 412. Dr. Warren, as Mofely juftly obferves, had fcarce any idea of this fever at all, except in its defcription. Towne calls it febris ardens biliofa, but does not add contagiofa, and afferts it to be an endemick of the Weft-Indies.

That it appeared, at Barbadoes, anterior to the time (1725) fixed by Dr. Warren is evident from Dr. Gamble, who well remembered it to be very fatal, in the ifland, in the year 1691. That Warren's account of the difeafe is altogether fabulous is certain ; this man, wrapt in hypothetical chaos, conftantly erred.

Pouppé Defportes fpeaking of the la maladie de Siam, fays ; " la régularité avec laquelle elle fe reproduit, femble devoir la faire regarder comme une de ces maladies dont il faut chercher la caufe dans la conftitution de l'air." Page 191, Tom. 1.

Dr. Cullen fuppofed this difeafe to arife from human effluvia and therefore, placing it under the fection of contagious difeafes, called it typhus Icterodes—but it is certain Dr. Cullen never faw the difeafe, and equally certain that he borrowed his idea from Dr. Warren—Dr. Cullen is confequently neuter on the prefent bufinefs.

As well might you fay, that a fkilful gardener could make a tree flourifh in a foil unnatural to its growth, as phyficians ingraft contagion on marfh exhalations.

In the almoft endlefs chain of cafes, wherein the clothes of thofe who have died, in the Weft-Indies, from the yellow fever, were brought to America, could contagion be explored but in one inftance,

even by the fagacious induftry of the indefatigable
Rufh. A gentleman's clothes being returned in
a trunk to his friends ; a young gentleman, upon
opening the trunk, became immediately unwel,
but no other perfon fuffered in the leaft. Thofe
clothes, beyond a queftion, were damp, this damp-
nefs was by heat converted into effluvia, corref-
ponding in every particular with marfh miafmata.
This effluvia, from the clofenefs of the trunk not
being able to efcape, remained in its offenfive and
concentrated ftate : And, upon the trunk being
opened, applied itfelf in its full force to the excit-
able fyftem of the young man. In this manner,
from the water in the clothes being changed into
marfh effluvia, and not from the clothes acting as
the vehicle of the human effluvia, was this gentleman difeafed. Now if the fuppofed contagion was
fo very vivacious as to perpetuate itfelf after fo
long a lapfe of time, how is it that the recent ema-
nations from his body treated with fuch great ten-
d rnefs the attending friends, as not to interfere in
the fmalleft degree with their healths ?

The idea of contagion is indirectly injurious to
commerce, and directly to fociety. Under the
influence of the perfuafion, that there is one of the
moft violent of contagious difeafes prevailing in feveral of the fea-ports of America, can we fuppofe
that foreign ports will fuffer our veffels to enter ?
A long and dangerous quarantine muft be performed ; the damages, accruing from fuch remora, the
merchants too fenfibly feel to be ignorant of. Even
when our papers do not echo the melancholy tid-
ings ; feeing that fuch difeafe has frequently vifited
our cities, is it not probable that they will guaran-
tee, their own fafety, by prohibiting the entrance
of our veffels? will they not naturally fay ? the
Americans have feen their error in rendering publick their difeafes—their filence is a piece of policy

—the difeafes exift.   Intereft, the main fpring of human action, forbids their publicity.

From flender caufes great events come to pafs. Few could have fuppofed that the intereft France took in the American ftruggle would have laid the corner ftone of her ruin and overthrow?—The government I fpeak of.   Athens was ruined folely ▉▉▉▉▉▉▉▉ giving countenance and fupport to a defpicable tribe of ftage-actors.   Upon a flender pivot play weighty matters.

Our extraneous interruptions are not the only inconveniences we labour under, domeftic commerce is fubject to arreftation, a general calm and ftagnation in bufinefs enfeebles our afcent.

Has it ever been known that the yellow fever has been propagated, through Baltimore, or any other city, from a perfon bringing from Philadelphia this malady, even in inftances where it has proven fatal ? The archives of America will never notice fuch a fact.

Do not our grave proclamations, and ferious refolves of city-corporations challenge our aftonifhment ? what are thefe proclamations and refolves for ? to prevent, what, from the earlieft dawn of lapfing ages has never, and to the lateft eve of expiring time will never, come to pafs.   Our intereft and commerce fall a facrifice, and are immolated at the fhrine of our folly.   As long as my reminifcence will afford me a knowledge of the proceedings of the health-committee of the city of Baltimore, I fhall admire and efteem their judgment and good fenfe.   Has the yellow fever ever been imported, from the Weft-Indies to America ? report fays it has, but where are proofs ? fmothered in murky obfcurity ; they fly the face and converfe of inveftigation.   Has this fever ever been carried from the Weft-Indies to Britain ? No.

E

The poor unfortunate fubjects of difeafe, flying
from the cities, find the doors and windows of the
country barred againft them.   The children leave
their parents to die, and parents their children,
their minds being joftled by the found of contagion, from their proper feats.   The lonely hearfe
folemnly conveys the dead to the dreary repofitory
of the filent multitude, where reigns ▬▬▬▬
and death.   Having confidered the hiftory and
nature of the yellow fever, we make a natural
tranfition to its fymptoms, firft premifing a few
pages on the proximate caufe.

CHAPTER II.

THE fabrick of the pathology of difeafes, has for more than two thoufand years floated on the varying ocean of incertitude, the fport of winds and tide. When the microfcopick eye traverfes the hemifphere of medicine, it beholds theories hurled on theories, fancies crufhed by fancies, and leffer errors fmothered by thofe of greater bulk and effrontery. From the aufpicious days of Hippocrates, we gently glide down the filent tide of time, collecting as we move the crazy wrecks of fhattered fyftems, until we arrive at the fluctuating variety of modern hypothefes. Hippocrates, wrapt up in the flattering pretenfions of his humoral pathology, and balancing between heat and bile, a long time fwayed the fceptre of the medical world. From an attentive perufal of this author's works, heat or bile, or plethora or obfruction (for in different paffages he fpeaks of all thefe) feem to conftitute the proximate caufe of fever. His fuccesfors foon perceived the futility of this foundation, and attempted to fabricate others more probable.

Diocles of Caryftus, a phyfician who flourifhed at an early period, and a man of confiderable eminence, afferted that fever was not fo much a primary difeafe, as fecondary and dependent on fome more hidden derangement. In order to avoid the force of his doctrine, after phyficians eftablifhed the diverfity of fymptomatick and idiopathick fevers. Prefently after Diocles, Erafiftratus, a phyfician at the court of Antigonus, invited the attention of the world, his proximate caufe refided in an error loci. Next Afclepiades, the Bythinian, ftepped on the ftage and rudely grafped the reins of control ; he, adopting the doctrine of atoms, hand- ed to the Greeks by Democritus of Abdera, at- tempts to account for the difference of types by a difference in the fize of the corpufcles, which he fuppofed to be formed by a combination of indivi- fible atoms. Here emerges the doctrine of the ob- ftruction of the permiable canals of the body, and its confequence modern vifcidity and lentor, fo fa- mous in the fchools. Afclepiades was the father and patron of the fect of the Methodicks.

Themifon vibrated between ftrictun et laxum, and on thofe two pillars reared his pathology of difeafes, here are the firft traces of fpafm, after- wards laboured by Hoffman and matured by the great Cullen : this hypothefis claimed the afcen- dency, at Rome, for more than an hundred years. At laft Galen, the impaffioned admirer of Hippo- crates, exhumed and reanimated the humoral er- rors.

Athenœus ventured to refufcitate the doctrine of the putrefcency of the blood (this is compre- hended in the writings of earlier authors) and pu- trefcency has not made a very defpicable figure in the world. Avicenna exprefsly defines fevers to arife from a preturnatural heat of the heart.

The Galenifts prevailed until the beginning of the fixteenth century ; about which time Aureolus Phillippus Theophraftus, commonly known by the name of Paracelfus, began to make a figure. This man affailed the Galenical party with all the en- gines of effrontery and fources of unimproved chy- miftry. Here commences the period of medical romance, fo fraught with the ftruggles between the mechanical and chymical modes of reafoning, thefe eventually neutralized, and ufhered in the chymi- co-mechanical philofophy.

The furious archeus of Van Helmont, differently modified, is the efforts of nature, fo celebrated
by Campanella and Sydenham, and autocrateia
of Stahl. If we except Mundy, Borelli and Cole
are the only writers previous to the time of Hoffman, who confidered the nervous fyftem as directly affording a feat for the proximate caufe of
fevers : Here in more pofitive terms originates
the fpafmodick ftricture. This idea of Hoffman,
Cullen has elaborated to its utmoft perfection.
Who has taken the proximate caufe from the heart,
and fixed it on the fuperficies of the body, in an
atony and fpafm of the capillaries.

After the above recital the mind is reftlefs, on
the poignant tenters of expectation, to embrace
a knowledge of the true nature and feat of the
proximate caufe. I fhall, leaving the anterior
whims to flumber with their authors, they being,
by one obliterating ftroke of the pen of Dr. Rufh,
fentenced to perpetual filence, take the liberty of
putting the opinion of the celebrated Philadelphia
Profeffor into the crucible of analitical inquiry.
The opinion of Doctor Rufh is the lateft that I
have met with in the writings of phyficians. His
words are " having premifed thefe general pro-
pofitions, I go on to remark, that a fever (when
not mifplaced) confifts in a morbid excitement
and irregular action in the blood-veffels, more efpecially in the arteries." " This irregular action
is in other words, a convulfion in the fanguiferous,
but more obvioufly, in the arterial fyftem." Page
134. Vol. 4. " From the facts and analogies
which have been mentioned, I have been led to
believe that irregular action or a convulfion in the
blood-veffels, is the proximate caufe of fever."
Pag. 139. Vol. 4.

There incontrovertably is a difference between
a fever and the proximate caufe of a fever; a fever

cannot confift in, or be made up of (thefe are
fynonimous) a convulfion in the blood-veffels, and
a convulfion in the blood-veffels be the proximate
caufe of a fever. An effect and the caufe of that
effect cannot be the fame. If the fever confifts
in, or is made up of, an irregular action or convulfion of the blood-veffel, what is the proximate
caufe? and vice verfa. It is illogical to identify
caufe and effect.

The proximate caufe, of which writers fay qud
prefens facit, fublata tollit, mutata mutat, after
being hunted from one part of the body to another,
and metamorphofed from one thing into another,
at laft takes refuge in nullibiety; having for its
affociates Phlogifton* and the four elements;†
and an honourable body they form, each having
enjoyed its apotheofis.

In a difeafe there are three effential caufes. The
predifpofing, or an aptitude of the body to be
acted on; the occafional, which acts directly or
indirectly on the feat of life and action, and the
exciting, or that which deftroys the equilibrium
of the nervous energy, and by this deftruction of
the equipoife of the fyftem gives the occafional
caufe an opportunity of victory. The caufation
or modus operandi which takes place between the
occafional caufe and the living principle is not ac-
ceffable even to the moft vigorous efforts of the human mind.

An offending entity (the occafional caufe) affails
the tranquility of the healthy body, a particular,
though infcrutable, infraction of the harmony of

---

* That accommodating nonentity of Stahl, which fo much
and fo long amufed and euled, one while under the form of
fixed fire, now in the fhape of fixed light, and anon in the
invefliture of the metallizing principle, the uncertain grafps of
infant philofophy.

† Thofe four peripatetick follies which for fo many hundred
years difgraced the human underflanding.

the animal œconomy is caufed, evidencing itfelf by a chain of fymptoms more or lefs unequivocal; thefe conftitute the fymptomata of writers, that (the primary derangement or infraction) the difeafe; here we obferve a regular and fimple concatenation of caufe and effect, and the evidences of fuch an effect. We cannot apply the name of difeafe to an arrangement of fymptoms, no more than we can the appellation of matter to an affemblage of qualities, or the epithet of fpirit to a combination of modes, but to that particularly morbid ftate of the body giving origin to fuch an arrangement of fymptoms.

Chills, fever, pain, proftration of ftrength, dif-colouration of the tongue, &c. are fymptomatick of a hidden and effential derangement (a difeafe) of the nervous power. Figure, divifibility, exten-fion, and folidity are indicative of an infcrutable material fubftratum. Paffion, memory, and judg-ment are proofs of an immaterial effence (afking Dr. Prieftley's pardon) the nature of which the labour of the human mind cannot develop.

The cherifhing beams of philofophy have begun to dawn, and I hope foon will enable us to proceed with more certain ftep in the healing art. It is time that we fhould divorce from our minds the pi-titio principii, and, like Pyrrho, difrobe ourfelves of credulous facility. This epoch demands felf-evident premifes or proven data for the ground-work of our inductions. Jurare in verba magiftri is the motto of unthinking hebetude, but a mafter's nod ought not to block up the avenues of refearch.

CHAPTER III.

D I A G N O S I S.

THAT affemblage of fymptoms, which generally are the appendages of any difeafe, and eftablifh
a barrier of divifion between it and all others, conftitute its pathognomonicks or diagnofticks. The
general charaƈteriftics, which disjoin the yellow fever from all others, are the following,

Its precurfers are, in fome inftances, a proftration
of fpirits and an inaptitude to motion, a fenfe of
uneafinefs and great fatigue ; pain and uneafinefs
through the limbs, as if from riding. It fome
times, without any premonition, impugns the
guardians of life. It will in one inftance affume
the drefs of the tertian, and in others clothe itfelf
in all the charaƈters of a cold. But let what will
be its harbingers, it foon hangs out its own colours,
and demands a tribute. The eyes become more
or lefs affeƈted by inflammation, accompanied with
an acrimonious or burning epiphora ; the head
feels itfelf molefted by pain and giddinefs, and a
fenfe of congeftion ; the tongue is indifferently
white, yellow, blue, red, brown or black ; in the
firft days of the difeafe it has an oily feel. A pyrexia attends ; the fkin is one while hot and dry,
at other times preturnaturally cold and clammy.
The præcordia is much oppreffed, attended by a
great inclination to vomit. Vomiting not unfrequently, or a cholera morbus or a diarrhœa gives
notice of the approaching calamity. The ftomach
in the latter ftages of the difeafe labours under a
fenfation of having in it fomething which it cannot

digeft; this fenfation they attribute to whatever they have fwallowed : A flatulency and hiccough help to fill up the train of evils. There is pretty uniformly a paucity of urine, and what is voided is very high coloured. A black vomiting or purging, hemorrhages from every part of the body, efpecially the ftomach, uterus, bowels, noftrils, and the incifions made by the lancet in bleeding; carbuncles and numerous little boils, more or lefs, act their part in this tragical fcene ; the black matter and hemorrhages feldom appear until after the fourth or fifth day ; yet they fometimes occur earlier. The countenance has a peculiarly ferocious look. The eyes are with the rednefs, tinged with a croceous colour ; this yellownefs frequently diffufes itfelf through the whole fuperficies of the body.

There is in fome cafes, about the fifth or fixth day, a ceffation of the fever, and all the violent fymptoms, every thing becomes apparently favourable, and the phyfician will augur aufpicioufly ; but this is a mournful circumftance ; it is the powers of life ceding, and not a relaxation of the difeafe ; efpecially if yellownefs and hemorrhages co-exift. There is frequently a metaftafis to the tefticles.

In puking the patient fome times throws up nothing but what is taken into the ftomach rendered a little ropy, at other times a black liquid refembling a mixture of foot and water is ejected.

The blood when abftracted is feldom covered with a buffy coat, but generally is what we call, a denfe, red blood ; it rarely is diffolved. An obmutefcence, and deafnefs are among the laft marks of an approching diffolution, they are truly prophetic of death. The fenfibility of the furface of the body is exceedingly morbid; and on

the leaft touch communicates uneafinefs ; this
preturnatural excitability I have met with in a fur-
prifing degree.

Not unufually a confiderable degree of delirium
accompanies this moft prominent in the black
catalogue of human ills. Perhaps there is no
complaint, from the effect of which patients are
fo long convalefcing. Small purple fpots very
often variegate the arms, breaft, and neck ; they
are ominous of peril.

By the diffection of defunct bodies, we get a
view of the dreadful ravages of this relentlefs
malady : We behold the ftomach disfigured with
fphacelated fpots, and characters of inflammation.
The liver fwollen, and exhibiting every mark of
phlegmafia : The fpleen preturnaturally flaccid :
The gall-bladder turgid with black and acrid bill ;
the whole of the primæ viæ, when a natural diarrhœa co-operates, is manifeftly affected with eryfi-
perlatous inflammation, which by the way is the
fpecies of inflammation that attacks the ftomach,
and this, probably, is reafon why the blood,
when abftracted, is not fizy ; were it of the
phlegmonoid fpecies the fize on the blood would
uniformly appear ; in fome cafes this fpecies of
inflammation does attend, and in thofe the pleuri-
tic coat, cæteris paribus, proclaims its prefence.
The eryfipelatous fpecies is generally too rapid in
its progrefs to mortification to communicate the
buff to the general mafs of blood. That the
buffy coat is an infeparable and infallible fign of
inflammation when accompanied by fever, and
vice verfa : See the experience of the moft en-
lightened and affiduous practicioners ; alfo the
profeffors of Edinburgh ; under whofe wings,
were folly and ignorance to deluge creation, learn-
ing and fcience would find fhelter.

The lungs not unfrequently fhow marks of

inflammation. The encephalon is pretty uniformly implicated in the teſtimony of the general deſtruction occaſioned by the yellow fever operating on the animal frame: Its meninges are found inflamed, the cortical and medullary ſubſtance itſelf is unuſually red.*

This diſeaſe attacks ſometimes with ſuch violence and ferocity, as, either from its force or the feeblenefs of the patients, to ſuperſede, by death, moſt of the above-mentioned appearances.

The black vomiting, hemorrhages, cerebral affections, yellownefs of the eyes and ſkin, purple ſpots, and pyrexia being the moſt conſpicuous and inſeparable diagnoſticks of the complaint, merit more ſerioufly our particular diſcuſſion.

The bile, in its natural and healthy ſtate, is as bland and mild as any ſecretion in the human body, but when the liver is affected by any ſpecifick action, its ſecretory function, like ſimilar phyſiological proceſſes, is ſubject to vitiation ; and that this black matter, diſcharged indifferently upwards or downwards, is vitiated bile, depending on a morbid action in the ſecretory organ, is obvious. 1. From the great quantity which is found upon diſſection, in the gall-bladder. 2. From its great acridity.* 3. From its analogy to other diſordered ſecretions : For inſtance that of the kidney in diabetes, of the ſtomach in diſpepſy, of a ſore when it becomes, as we generally phraſe it, vitiated. Laudable pus is a ſecretion, ſo is the acrimony eſcaping from a vitiated, a cancerous, or a ſcrofulous ulcer. The atrabilis of the ancients perfectly accords with the black vomiting of the moderns. Some phyſicians have perfuaded themſelves that this black matter is owing to an admixture of blood, and that there is an abſolute want of ſecretion in the yellow fever ; this opinion, I

* Vide Pouppé Deſportes, Hillary, Moſely, Rufh, &c. &c.

muſt believe, has a more intimate affinity with prejudice, than with reflection.

Hemorrhages ſeldom overtake the patient in the early ſtage of the diſeaſe, except under the form of epiſtaxis, which is not commonly conſiderable. When they do occur, they are the indubitable evidence of an atony or paralyſis of the blood-veſſels and this atony in its turn depends on a deſtroyed vigour of the nervous power ; this deſtruction of the nervous influence is effectuated by the peccant agency operating directly and immediately on the ſenforium commune. It is the reſult of a violent action on the immediate ſeat of life, and not of the vaſcular, or arterial ſyſtem ; the blood-veſſels can only be acted on ſecondarily. If we could timouſly diminiſh the quantity of the hoſtile entity, we ſhould infallibly prevent thoſe ſanguineous fluxes. Moderate blood-letting, by leſſening the volume of the blood, will alſo contribute conſiderably to that end ; but too profuſe bleedings, by robbing the ſyſtem of its ſources of recovery from ſo violent a ſhock, precipitate the unhappy ſufferer to his grave. That a violent action on, and not a ſmothered or incontrolable exertion of the ſyſtem, lays the foundation of thoſe hemorrhages, is patent. 1. As they do not accompany phrenitis or pneumonia where the vaſcular exertion is much more fierce, coercive, and oppreſſed. 2. Blood-letting after the ſecond or third time, except in particular habits, when it is pretty copious, ſo far from relieving the imaginarily oppreſſed pulſe, rather renders it more feeble and yielding. 3. In the moſt violent attacks, where there exiſts the greateſt degree of indirect debility, the propriety and ſafety of immoderate phlebotomy are in the inverſe ratio of this indirect debility. I ſpeak from my own experience, and that of Hillary, and Pouppé Deſportes who carried the lancet to its greateſt extremes; and poſſeſſed the moſt ample opportunities of ſeeing its advantages and diſadvantages; not for one ſeaſon only but for fifteen or twenty years, during which time he practiced in the Weſt-Indies; as alſo of ſeveral of the moſt reſpectable phyſicians in the city of Baltimore. Laſtly hemorrhages are more apt to occur in thoſe who have been copiouſly, than in thoſe who have been moderately bled; provided untimely death, ſuperinduced by thoſe large evacuations, does not obviate them. So is mortification about the orifices made by the lancet. Pouppé Deſportes, ſpeaking of thoſe hemorrhages, and mortification, remarks; " Dans pluſieurs les ſaignées ſe rouvient, et le ſang, malgré le nombre des compreſſes, pénetre, cette hémorragie eſt ſouvent accompanée d'une gangrene charbonnée, qui ſe forme autour de la ſaignée, et dont on ne peut, arrêter le progrès." And a little below.— " Cet accident arrive ordinairement à ceux qui ont été trop ſaignés."* I ſhall ſpeak more diffuſely on this when I arrive at the therapeutick diviſion of this publication.

The Encephalon, although ſubject to inflammation and partial infarction, cannot labour under general congeſtion; that is, the brain-caſe cannot contain more at one time than another; except before the bones and ſutures become firm and oſſified, and in caſes where the ſutures are afterwards deſtroyed by diſeaſes. The medullary ſubſtance of the brain is incompreſſible,† and the caſe itſelf is unyielding: When a congeſtion of the right ſide of the head happens, there is a ſimultaneous, and equivalent diminution of the ariæ of the veſſels of the left. In caſe of hydrocephalus there is a general invaſion of the ariæ of the cere-

* Vide Tom. 1. p. 200.
† Vide profeſſor Monro's incontrovertable experiments.

bral veffels. In fome inftances the cortical fub-
ftance is worn away by attrition. If a general
congeftion of the brain could take place, a partial
vacuum would of confequence be poffible ; and if a
partial vacuum took place, the plates which are be-
hind the eye-bales would unavoidably be forced in,
and inftant death be the refult by the weight of the
external air :   The columns of which, in weight
amounts to about forty-two pounds on each eye,
allowing fourteen pounds to each fquare inch.*
The pofterior plate that fupports the eye is frequently fo thin as to be quite diaphanous. The
inference is that either an inflammatory or partially congefted ftate of the brain muft give fupport
to that delirium, vigilance &c. we obferve in the
yellow fever and many other complaints.

This receives acceffion of certitude from the
circumftances attending the decapitation of an
animal ; when the head of an animal is fevered
from its body, all the blood difcharged is from the
external parts of the head, not one drop efcapes
from the internal. This is fubftantiated by the
following experiments. Diffect the veffels to their
exit from the fkull, then fecure them firmly by a
ligature, this being done, divide the cafe and you
will find every veffel regularly filled and replete
with its contents having fuffered no evacuation.

Again, take a glafs-globe with two oppofite
orifices, fill it with water, then put your finger
upon one, and turn the other downwards ; no
water will efcape until you remove your finger
from the fuperior orifice. Nor can the contents of
the fkull be either decreafed or augmented, except the cafe previoufly be expofed to violence.

The yellow or brown colour of the eyes and
fkin is owing to an abforption of the bile or brown
matter after it is fecreted. There is fometimes a

tempory yellownefs of the fkin, this may be produced by a peculiar action of the blood-veffels.*

The purple fpots conftitute a pathognomonick,
they are neither the production of an over action
of the fyftem, nor a diffolved ftate of the blood.
They are to the yellow fever what the red fpots
are to the cynanche maligna, or the eruptions of
the fkin to the meafles ; they are fymptomatick,
and no regular confequence of the general ftate
of the fluids or condition of the folids.

Pyrexia or fever, as afferted by Dr. Clark,† and
afterwards by profeffor Rufh, is fimple, and has
no generick difference ; there is but one fever :
A fever is the mode of a difeafe, and not idiopa-
thically the defeafe.‡ It, with other concomitant
qualities, evidences an infraction of the harmony
of the animal œconomy. Extenfion and figure
are the qualities of matter ; thought and reminifcence the modes of fpirit.   :

Small pox propagate fmall pox, and meafles the
meafles : The fmall pox and meafles are generically different ; a fever accompanies both : If the
fever were idiopathick in either, a fever would
poffefs generick difference. A fever is fimple or
an unit, the corollary is then, a fever is the mode
or quality of a difeafe.

Fever is a convulfive action of the arterial fyftem, as mentioned by Dr. Rufh, accompanied
by more or lefs of a peculiar and indefcribable
heat and drynefs of the fkin : The drynefs and
heat though are not always prefent ; as in the
febris typhodes. This convulfion of the arterial
fyftem is the refult of a more infcrutable and hidden morbidity of the fource and feat of life and
action, the nervous power. The mufcular fibre

* Vide Rufh.
† Vide medical commentaries by Dr. Duncar.
‡ Vide Diocles.

poſſeſſes no vis inſita, its ſenſibility and contractility are feudatory of the vis nervia.*

The ſource of life and action muſt be originally and primarily concerned in all the operations, whether morbid or healthy, of the animal body. The nerves are the " ſeat and throne" of all diſeaſes.

Whether a fever be the reſult of a mere mechanical action, or the provident effort of the vis medicatrix naturæ, that power in the body to heal its own maladies, I leave to the more learned to determine. But there is a ſomething about a fever which is eaſier recognized than developed.

A peccant entity, acting either mediately as in the form of fracture or wound, produces fever indirectly ; or immediately as in the ſhape of human or marſh effluvia &c. produces fever directly. In both caſes the fever is the ſame, and ſymptomatick : In neither is it idiopathick. It is nothing but a ſymptom of a more primary and eſſential derangement of the nervous energy.— This primary and eſſential derangement is uniformly the diſeaſe ; whatever ſucceeds is only indicative of this firſt diſturbance of the quiet of the animal ſyſtem.

* Vide Monro's nervous ſyſtem.

CHAPTER IV.

C U R E.

WE now arrive at the divifion of this effay
which is the moft interefting and merits the moft
ferious and unbiafed inveftigation. And leaving
the flippery declivity of hypothetical change, we
introduce our readers to the more unequivocal and
inflexible data of practical experience : where feeble theory is fupplanted by more certain practice,
where the fick bed triumphs over the reveries of
the clofet.

Dr. Rufh, in manfully and fuccefsfully labouring to ftem the torrent of error and prepofterous
madnefs which had diffufed themfelves throughout,
and woven the tiffue of general practice ; and call-
ing the medical mind back to the almoft antiquat-
ed fyftem of depletion fo fortunately purfued by
Sydenham, Cullen, Monro, Gregory, Botallus,
Pouppé Defportes &c. has attached immortal honour to himfelf, and, ufing a gallicifm, deferved
well of mankind. The doctor's exalted dignity
elevated him above the mean wiles of plagiarifm.
He, with the candour proper to great minds, frank-
ly acknowledges his obligations to preceding writers. I am though confiderably perfuaded that in
his ftrenuous exertion to crufh the growing folly
of medical prejudice, which took root in the exe-
crable writings of Brown of Edinburgh, and Kirk-
land of England, he has fuffered himfelf to be hur-
ried within the embraces of the oppofite extreme.
It is beyond contradiction that the experience of

fome of the moſt ſcientific and beſt informed phyſicians, will not warrant the extremes of depletion
inculcated in his learned works.

In the management of the autumnal remittert
or yellow fever there are four therapeutic inten-
tions.

1. To diminiſh the violent action of the general
fyſtem and remove as far as poſſible the inflammatory difpoſition of the liver, ſtomach, &c.

2. To take off the ſtricture of the ſuperficies of
the body.

3. To difcharge the acrid bile as quickly as it is
excreted.

. 4. To reſtore the vigour of the frame as foon as
poſſible after the fever has fubfided.

The firſt intention is beſt accompliſhed by a ju-
dicious and proper ufe of the lancet, together with
a fpeedy introduction of mercury, either in the form
of calomel or ointment, into the fyſtem. It will
too often happen from an over excitability of the
ſtomach that the calomel cannot be ufed. When
this occurs the mercurial ointment muſt be freely
applied to the infides of the thighs, legs, and arms ;
thofe parts being the moſt abundantly, of all the
external parts of the body, fupplied with the lym-
phatics.

Calomel is the moſt efficacious and powerful of
all medicines in the refolution of inflammations of
whatever kind they may be. But in thofe of the
liver its falutary effects are peculiarly deferving our
notice. For this valuable information we are principally indebted to Dr. Gilchriſt of Scotland and
the practitioners of the Eaſt-Indies.

- When we enter the room of a patient in this fever our firſt attention ſhould be to the ſtate of the
eyes, the degree of pain in the head, oppreſſion a-
bout the præcordia, and fullneſs of the pulfe. This
is feldom or never hard ; indeed the ſtroke of the

artery is more deceitful in this fever, than in any
other difeafe I have ever met with. If the eyes be
much inflamed, or labour under a fenfe of protru-
fion from the fockets, or an unweldinefs in their motion : an acid epiphora frequently diftils from the
eyes ; together with the above if the head com-
plains much, the pulfe appear full to the application of the fingers, and be frequent ; we muft in-
ftantly have recourfe to the lancet. Even fhould
the tongue at the fame time be blue or brown.
The colour of the tongue is not to be in general
confided in. It is fometimes blue or brown from
the firft days of attack. Whilft ever the pain of
the head, or back continues, with a very frequent
or full pulfe the lancet muft be reforted to. This
though will rarely be the cafe after two or three
good bleedings, except in particular habits. Some
habits will bear the lancet to the fifth or fixth re-
petition, efpecially where the indirect debility of the
fyftem is not great ; where there is great heat of
the furface of the body, and the tongue white.
A timous ufe of the lancet more effectually than
any other remedy, tends to prevent the rapid progrefs of the inflammation of the different vifcera.
When it has taken place blood-letting promifes
more liberally than all other remedies. But when
bleeding is carried to too great extremes, it exhaufts
the general fyftem and proftrates the powers of
life in fo great a degree, that the animal frame can
never renew its functions ; it haftens, by robbing
the blood-veffels of internal fupport and nourifh-
ment, that atony and palfy of the vafcular fyftem
which lays the ground-work of thofe melancholy
hemorrhages.
When blood-letting is had recourfe to, it fhould
be practiced within the firft three or four days. It
may, under particular circumftances, be performed
at later periods, but not with fuch propitious confe-

quences. I have practiced it, as late as the tenth and fifteenth day of the difeafe, in inftances where the patient's ftrength had not been fapped by evacuations.

Experience together with the writings of moft of the refpectable practitioners of the Weft-Indies eftablifh the following,

Thofe who were not bled fuffered from the neglect of the lancet.

Thofe who were bled largely died from the abufe of the lancet.

Thofe who were bled according to the pain of the head, fullnefs of the pulfe, and oppreflion about the præcordia, or in other words moderately, in a much greater proportion recovered. Bleeding is by no means new in the yellow fever.

There were two young gentlemen, who in a vifit to the Point contracted the difeafe: They were both taken ill a few day after their return to the Town. One of them was blooded fix or feven times, and died. The other was bled once only and recovered. The violence of the attacks was apparently equal.

A little after, there were two others who took the difeafe by going to the contaminated atmofphere ; one of whom was bled fix or feven times within about forty-eight hours, was put into the cold bath, had injections of cold water ; he died, amidft the hands of the operators, during the third injection of cold water. The other who was delirious almoft from the firft onfet of the complaint, was not bled at all, yet after a fevere and dangerous illnefs recovered.

There were two other cafes, one of whom loft about as much as would be taken away at four common bleedings, and recovered. Though no patient could be worfe than fhe was to recover.— The other was bled twelve or thirteen times, his arm mortified and he departed from among the living. Out of fix who took the difeafe at or near Bowley's wharf, five, fome of whom were fo much evacuated that their friends threatened to interfere, died. The one, Mr. Waters, who efcaped was bled but twice and that moderately. There were a few inftances of recovery after thofe profufe evacuations, but they were relatively few. Where one, after fuch immenfe loffes of blood efcaped, there were ten who were either not bled at all, or but once or twice, that recovered. Out of all the blacks, for negroes by no means efcaped this complaint, whom I attended from the firft attack of the difeafe, and they were not few in number I loft none. I bled but one, and him only to the amount of fix ounces. I do not recollect that I met with a fingle inftance of hemorrhage in a black perfon.

Calomel not only is the moft effectual medicine that can be ufed in the firft ftage, but is alfo the only one in which we can have confidence to remove the ftricture of the furface of the body; efpecially when affifted by the warm bath, either generally or partially applied. I have feen the ephidrofis produced to that extent by a liberal ufe of calomel, as to require three or four changes of linen in twenty four hours.

Calomel not only cures by acting as a diaphoretic and antiphlogiftic, but alfo by eftablifhing in the fyftem an oppofite action to that of the fever. No two general actions can exift at the fame time: So that when the mercurial action takes place the morbid one muft, of neceffity, ceafe. The eftablifhment of this mercurial action is confirmed to the practitioner by a free ptyalifm. Whenever a free falivation takes place the patient is fafe. Perhaps no perfon ever died after the full eftablifhment of this difcharge from the gums. Not that the falivation, ſtrictly ſpeaking, is of any ſervice in itſelf. It is in the yellow fever, as in the lues venerea, the unavoidable conſequence of the general mercurial action of the ſyſtem, and of no farther ſervice than informing the phyſician of this general action.

That the local pain is of no advantage is evident from the following : Let the gums become ever ſo much inflamed, pained, and ſwollen, if a very free ſpitting ſhould not ſucceed, the ſick reaps no advantage, but on the contrary this ſtate of the gums is ominous of approaching death.

To invite the mercury to the ſurface of the body, the tepid bath ſhould be uſed, or in its place the pediluvium, and local applications of flannel, wrung out of hot water, to the regions of the liver, ſtomach, &c. Thoſe ſhould be frequently repeated, and continued, at leaſt, half an hour each time.

The cold bath has been very ſtrenuouſly recommend by Jackſon. But it does not appear from his writings that he ever uſed it with ſucceſs without its being preceded by the warm bath : This together with its ill effects whenever it has been introduced into practice in America,* render it probable that Jackſon attributed to the cold, what belonged to the warm bath. If it has ever been of the ſlighteſt utility in caſes in America, theſe caſes have not as yet found their way to the public. I ſay from the cold bath, in the practice of Dr. Jackſon, being uniformly preceded by the warm, and its general failure in America when uſed alone, I am fully perſuaded that it not only is a uſeleſs, but a dangerous application in the yellow fever.

Emetics even of the gentleſt kind are inadmiſ-

* Vide Ruſh.

fible in every stage of the fever.* No preparation of antimony, from the tendency in those articles to excite vomiting, can be used with safety. The justly celebrated James' powder is here of too precarious operation. All neutral salts are too inert and uncertain.

Whether, upon being called to the patient, we bleed, or not; we must instantly order a mercurial purge, which should be repeated every day or every second day, so as to produce four or five stools daily, this number he at least ought to have. To assist the operation of the cathartics; the lower part of the intestines must be opened by means of glysters. A glyster-syringe will with great propriety and utility be kept constantly in the patients room, that by the use of this the purges when they are too slow may be quickened and invited downwards. For as above mentioned the sick should never have less than four or five passages a day during the strength of the fever. Cathartics are of all remedies the best and most useful. They are not be omitted even in cases where the pulse is feeble, and intermitting : If purges cannot relieve the patient, his chance is truly melancholy.

In support of the great necessity of constant purging consult Mofeley, Hillary, Pouppé Desports, and the learned Rush. I had my patients to call for the close-chair from three to ten and fifteen times a day. And those who purged most, when not accompanied by puking, recovered soonest. In one case which I was called to, four grains of calomel produced thirty-five stools : after which evacuation the patient began to mend and recovered. A patient, to whom I was called, took a little broth which rendered him much worse than he had been the preceding day. I ordered him

* Some practitioners have given emetics in the instant of attack with supposed advantage.

thirty grains of jallap and ten of calomel (my usual dose) which operated twelve times; and from that instant he began to recover. It was to the repeated use of cathartics, and mercurial diaphoretics that I trusted the blacks to whom I was called.

Of all cathartics, I prefer gamboge and calomel, or jallap and calomel. By the timous use of the former of these I prevented the regular course of this fever, in at least thirty patients, who upon the first appearance of the disease called on me.

The fourth and last intention is employed principally in selecting those articles which will the most readily restore the exhausted strength of the patient. From habit and prejudice in this stage of the disease, physicians generally fly to the well known powers of bark and wine, opium and æther, colombo, and quassia, &c. But although these medicines are in appearance our hope and reliance; a just pathology and experience will quickly evince the impropriety and hurtful tendency of all tonic and stimulant remedies. The snake-root itself is too great a bitter, and must not, except under particular circumstances, be exhibited. Our sole trust and dernier resource is in a well regulated dietetic plan, with now and then a purge to keep the bowels open. Nor must the inexperienced suffer the weakness of the patient to frighten them from the free use of purgatives.

The habits and desires of the stomach are chiefly to be consulted after the fever has subsided. But during the heighth of the fever, neither flesh nor any thing made of flesh can be allowed. Vegetables are the only articles of food that the patient can be indulged in while the fever possesses any considerable strength.

The patient must never take any beverage stronger than barley-water, lemon-ade, cold water, water with a toaſt in it, and ſuch like mild potables. Some phyſicians are too cautious in giving cold water, and lemonade during the uſe of mercury. But I am ſatisfied from daily practice that no drink is more innocent and beneficial in a fever, eſpecially the yellow fever, than lemonade, and cold water : nor are we to diſcontinue the citric acid even during the uſe of calomel. Chymical affinities might induce us to believe that upon an union of the acid and calomel, the latter would, by decompounding the former, and attaching to itſelf part of its oxigen, convert itſelf into the oxiginated muriate of mercury, or what is vulgarly called corroſive ſublimate. This though in fact does not happen. If any acid poſſeſſing a greater affinity, to the baſe of calomel, than the muriatic, be uſed, the muriatic acid will be precipitated, and a new neutral, formed by this more powerful acid, and the baſe of the calomel viz. mercury, will be the conſequence.

Bliſters are ſometimes uſeful ; particularly when applied to the epigaſtric region. They quiet the diſturbance and excitability of the ſtomach. They are not unuſually applied to the extremities and other parts of the body, but perhaps more from cuſtom and faſhion than conviction of their real uſe.

Opium may, in the laſt ſtages when there is no fever preſent, be given in very ſmall quantities, thereby to take off for a few hours the great irritability of the ſtomach. But all opiates muſt be followed, within ſix hours after they are given, by cathartics ; otherwiſe their ſtimulant qualities will far more than counterbalance any advantages the patient may at firſt receive from them.

During the moribund ſtate of a patient, I have made uſe of æther and muſk, but my views con-

H

templated the removal of particular fymptoms,
fuch as hiccough, twitches of the tendons &c. which
very much diftreffed what few friends there were
who had courage enough to come near the depart-
ing fufferer ; and by no means the cure of the
complaint. They ferve under fuch circumftances
to fmooth the paffage to the grave, they cannot de-
liver from the pounce of the fell malady.

When hemorrhages do overtake the unhappy
fufferer, general experience has painfully convinced
us of the impoffibility of managing them by aftrin-
gents whether internal or external, except in thofe
of the uterus where applications of cold vinegar
for the moft part anfwer. We are here ftill to pur-
fue our general plan.

All applications to the carbuncles which oft ac-
comany this fever, are ufelefs. They may do mifchief but we can expect but little benefit from them.
Nor does opening of them feem to anfwer any
good end. It is beft to leave them to nature.

Appendix.

DURING the time that the foregoing Tractate was at the press, I accidentally met with the answer of the Physicians of Philadelphia, to the request of the governor, the hon. Thomas Mifflin—in relation to the yellow fever. A perfect and harmonious coincidence unites our sentiments, so far as they respect the origin and cure. For the satisfaction and benefit of the public, I will insert the letter.

SIR,

"IN compliance with your request, the subscribers have devoted themselves to the investigation of the origin, progress, and nature of the fever which lately prevailed in our city; and we have now the honour of communicating to you the result of our enquiries and observations.

We conceive the fever which has lately prevailed in our city, commonly called the yellow fever, to be the bilious remitting fever of warm climates, excited to a higher degree of malignity by circumstances to be mentioned hereafter.

Our reasons for this opinion are as follows :

1st. The sameness of their origin, both being the offspring of putrefaction. Of this there are many proofs in the histories of the yellow fever in the West-Indies. Where there are no putrefaction the West-India islands enjoy a perfect exemption from that disease in common with nothern climates.

2d. The yellow fever makes its appearance in these months chiefly in which the bilious fever prevails in our country and is uniformly checked and destroyed by the same causes ; viz. heavy rains and frosts.

3d. The symptoms of the bilious fever are the same in their nature. They differ only in their degree. It is no objection to this assertion that there is sometimes a deficiency or absence of bile in the yellow fever. This symptom is the effect only of a torpid state of the liver, produced by the greater force of the disease acting upon that part of the body. By means of depleting remedies this torpor is removed and the disease thereby made to assume its original and simple bilious character.

4th. The common bilious and yellow fever often run into each other. By depleting remedies the most malignant yellow fever may be changed into a common bilious fever and by tonic remedies, improperly applied, the common bilious fever may be made to affume the fymptoms of the malignant yellow fever.

*5th. The common bilious and yellow fevers are alike contagious, under certain circumftances of the weather and of predifpofition in the body. That the common bilious fever is contagious we affert from the obfervations of fome of us, and from the authority of many Phyficians who have long commanded the higheft refpect in medicine.

6th. The yellow and mild bilious fevers mutually propagate each other. We concieve a belief in the unity of thefe two ftates of fever to be deeply interefting to humanity, inafmuch as it may lead patients to an early application for medical aid, and phyficians to the ufe of the fame remedies for each of them, varying thofe remedies only according to the force of the diforder. It is no objection to this opinion that, that ftate of bilious fever called the yellow fever, is a *modern* appearance in our country. From certain revolutions in the atmofphere as yet obferved only but not accounted for by Phyficians, difeafes have in all ages and countries alternately rifen and fallen in their force and danger. At prefent a conftitution of the atmofphere prevails in the United States which difpofes to fever of a highly inflammatory character. It began in the year 1793. Its duration in other countries has been from one to fifty years. It is not peculiar to the common bilious fever to have put on more inflammatory

* This argument, the fpirit of which ftates this and the bilious fever to be contagious, appears to be the offspring of conjecture and a biafed education. Whatever the exertions and perfpicacity of fome individuals of Philadelphia may have explored I cannot determine. But, in my candid opinion, the obfervations, which deceived them into the perfuafion that the bilious fever is infectious, are, like Dr. Prieftley's chymical experiments, fomewhat defective. Many phyficians of eminence have afferted, in their writings, the above fever to be contagious ; but they have neglected to accompany thofe conjectures with tributary facts to fupport and give life to them.

There is not, fo far as my reading extends, one folitary cafe of a contagious bilious fever in the records of medicine : I fay not one cafe wherein contagion has been, or can be proved. And here, reader, my opinion is propugned and countenanced by Cullen, Sydenham and many more of the moft learned and experienced of phyficians. Bilious fevers fpread by what thofe learned gentlemen in the latter part of the following paragraph call " a conftitution of the atmofphere which prevails in the United States and difpofes to a fever (the bilious or yellow fever) of a highly inflammatory character. It is this conftitution, effectuated by the marfh effluvia or vegetable putrefaction being diffufed through the air, which in the firft place gives rife to, and afterwards renders epidemick or general the bilious or yellow fever, as thofe gentlemen in what follows of their reply wifely allow. Thofe who wifh to be fatisfied that this fever is not contagious, will do well to pay particular attention to the cafes inftanced in the above letter.

fymptoms than in former years. There is fcarcely a difeafe which has not been affected in a fimilar way by the late change in our atmofphere and that does not call for a greater force of depleting remedies than were required to cure them before the year 1793.

7th. And laftly. The yellow fever affects the fyftem more than once, in common with the bilious fever. Of this there were many inftances during the prevalence of our late epidemic.

The fever which lately prevailed in our city appears from the documents which accompany this letter to have been derived from the following fources.

1ft. Putrid exhalations from the gutters, ftreets, ponds and marfhy grounds in the neighbourhood of the city. From fome one of thefe fources we derive a cafe attended by Dr. Caldwell on the 9th of June—one attended by Dr. Pafcalis on the 22d of July, and by two cafes attended by Dr. Rufh and Dr. Phyfic on the 5th and 15th of the fame month; and alfo moft of thofe cafes of yellow fever, which appeared in the northern parts of the city, and near Kenfington bridge, in the months of Auguft, September and October. We are the more fatisfied of the truth of this fource of the fever from the numerous accounts we have received of the prevalence of the fame fever and from the fame caufes, during the late autumn in New-York and in various parts of New-Jerfey, Pennfylvania, Maryland, Virginia and South Carolina, not only in fea ports but inland towns. The peculiar difpofition of thefe exhalations to produce difeafe and death was evinced early in the feafon by the mortality which prevailed among the cats and during every part of the feafon by the mortality which prevailed in many parts of our country among horfes. The difeafe which proved fo fatal to the latter animals is known among the farmers by the name of the yellow water. We conceive it to be a modification of the yellow fever.

2ndly. A fecond fource of our late fever appears to have been derived from the noxious air emitted from the hold of the fnow Navigation, capt. Linftroom, which arrived with a healthy crew, from Marfeilles on the 25th of July, and difcharged her cargo at Latimer's wharf, after a paffage of eighty days. We are led to afcribe the principal part of the difeafe which prevailed in the fouth end of the city to this noxious air, and that for the following reafons.

1ft. The fever appeared firft on board this veffel and in its neighbourhood, affecting a great number of perfons nearly at the fame time, and fo remote from each other that it could not be propagated by contagion.

2dly. There was in the hold of this veffel a quantity of vegetable matters, fuch as prunes, almonds, olives, capets, and feveral other articles, fome of which were in a ftate of putrefaction.

3dly. A moſt offenſive ſmell was emitted from this veſſel after ſhe had diſcharged her cargo, which was perceived by perſons ſeveral hundred feet from the wharf where ſhe was moored.

4thly. A ſimilar fever has been produced from ſimilar cauſes, in a variety of inſtances : we ſhall briefly mention a few of them.

At Tortola a fever was produced in the month of June, in the year 1787, on board the ſhip Britannia, capt. James Welch, from the noxious air generated from a few buſhels of potatoes, which deſtroyed the captain, mate, and moſt of the crew, in a few days.

Two ſailors were affected with a malignant fever, on board the ———, capt. Thomas Egger, in the month of March, 1797, from the noxious air produced by wine that had putrified in the hold of the ſhip, one of whom died ſoon after her arrival in Phi-ladelphia.

In the month of June, 1793, the yellow fever was generated by the noxious air of ſome rotten bags of pepper on board a French Indiaman, which was carried into the port of Bridgetown, by the Britiſh letter of marque Pilgrim. All the white men and moſt of the negroes employed in removing this pepper, periſhed with the yellow fever, and the foul atmoſphere affected the town, where it proved fatal to many of the inhabitants.

On board the Buſbridge Indiaman a yellow fever was pro-duced in the month of May, 1792, on her paſſage from England to Madras, which affected above two hundred of the crew. It was ſuppoſed to be derived from infection, but many circumſtanc-es concur to make it probable that it was derived from noxious air. The abſence of ſmell in the air does not militate againſt this opinion, for there are many proofs of the moſt malignant fe-vers being brought on by airs which produced no impreſſion on the ſenſe of ſmelling. This is more frequently the caſe when the impure air has paſſed a conſiderable diſtance from its ſource and becomes diluted with the purer air of the atmoſphere.

Several caſes are related by Dr. Lind, in his treatiſe upon fe-ver and infection of the yellow fever, originating at ſea under circumſtances which forbade the ſuſpicion of infection, and which can only be aſcribed to the impure air generated from pu-trid vegetables.

So well known, and ſo generally admitted is this ſource of yellow fever in warm climates, that Dr. Shannon, a late writer upon the means of preventing the diſeaſes of warm climates, in enumerating its various cauſes, expreſsly mentions " the putrid effluvia of a ſhip's hold."

We wiſh due attention to be paid to theſe facts, not only be-cauſe they lead to the certain means of preventing one of the ſources of this fever, but becauſe they explain the reaſons why ſailors are ſo often its firſt victims, and why from this circum-ſtance the origin of the diſeaſe has been ſo haſtily, but erroneouſ-ly aſcribed ſolely to importation.

The fever which prevailed along the shore of the Delaware, in Kensington, and which proved fatal to Mr. Joseph Bowers and two of his family, we believe originated from the noxious air emitted from the hold of the ship Huldah; capt. Wm. Warner. This air was generated by the putrefaction of coffee which had remained there during her voyage from Philadelphia to Hamburgh, and back again.

* In the course of our enquiries we were led to suspect one source of our late fever to be of foreign origin. The sails of the armed ship Hinde, on board of which several persons had died of the yellow fever, on her passage from Port-au-Prince, and which arrived on the 4th of August, were sent to the said store of Mr. Moyse. Four persons belonging to the loft were soon afterwards affected with symptoms of a bilious yellow fever. We shall not decide positively upon the origin of the fever in these cases ; but the following facts render it probable that it was not derived from the persons who had died of it on board the suspected vessel. 1st, The sails emitted an offensive smell ; 2d, three of the cases of the persons affected in the sail loft were of a mild grade of the fever ; 3d, the fever was not propagated by contagion from any one of them ; 4th, the sail loft was within the influence of the noxious air which was emitted from the hold of the snow Navigation, being not more than fifty yards, and was in the direction of the wind which blew at that time over her. The extent of this air has not been accurately afcertained, but many analogies give us reason to believe that it may be conveyed by the wind, in its deleterious state, from half a mile to a mile.

In support of the opinion we have delivered of the origin of our late fever, we must add further, that in that part of the city which lies between Walnut and Vine-streets, and which appeared to be free from the effects of exhalation and the noxious air of the ships, there were but fews cases of the fever which appeared

* Relatively to the persons affected after receiving the sails of the armed ship Hinde, there is not the smallest probability that they were infected by the sails acting as fomites. The explanation given by those gentlemen is perfectly satisfactory. And had not the snow Navigation been in the vicinity, we should not have been at a loss to account for their indisposition without admitting the idea of contagion. The sails emitting an offensive smell is an irrefragable proof that they were damp; and what must be the state of the water and deleterious gas which were enveloped in their folds ? this putrid state of water is the very postulate to render it destructive.

It is impossible that Philadelphia amidst the numerous sources of vegetable exhalations could have one street or lane free of this miasma. And the sparse cases which happened between Walnut and Pine-streets were produced by a lesser quantity of this vegetable effluvia.

There is not in this nor any other writings of the physicians of Philadelphia a single case accompanied by even the probability of contagion. And there is no axiom more evident than that the bilious or yellow fever is not contagious.

to fpread by contagion, even under the moſt favourable circumſtances for that purpofe.

Having pointed out the nature and origin of our late fever,
we hope we ſhall be excufed in mentioning the means of preventing it in future.————Thefe are,

* Firſt. A continuance of the prefent laws for preventing the
importation of the difeafe from the Weſt-Indies, and other parts
of the world where it ufually prevails.

Secondly. Removing all thofe matters from our ſtreets, gutters, cellars, gardens, yards, ſtores, vaults, ponds, &c. which
by putrefaction in warm weather afford the moſt frequent remote
caufe of the difeafe, in all countries. For this purpofe we recommend the appointment of a certain number of phyficians
whofe bufinefs it ſhall be to infpect all fuch places in the city, the
northern liberties and Southwark, as contain any matters capa-
ble by putrefaction of producing the difeafe and to have them
removed.

Thirdly. We earneſtly recommend the frequent waſhing of
all impure parts of the city in warm and dry weather, by means
of the pumps, until the water of the Scuylkill can be made to waſh
all the ſtreets of the city ; a meafure which we conceive promifes
to our citizens the moſt durable exemption from bilious fevers of
all kinds, of domeſtic origin.

Fourthly. To guard againſt the frequent fource of yellow fever from the noxious air of the holds of ſhips, we recommend
the unlading all ſhips with cargoes liable to putrefaction at a
diſtance from the city, during the months of June, July, Auguſt,
September and October. To prevent the generation of noxious
air in the ſhips, we conceive every veſfel ſhould be obliged by
law to carry and ufe a ventilator, and we recommend in a particular manner the one lately contrived by Mr. Benjamin
Wynkoop.—We believe this invention to be one of the moſt
important and ufeful that has been made in modern times, and
that it is calculated to prevent not only the decay of ſhips and
cargoes, but a very frequent fource of peſtilential difeafes of all
kinds in commercial cities.

* This difeafe can only be imported from the Weſt-Indies or any
place by veſfels arriving from thofe places with putrid water or vegetables in their holds, &c. As this fever uniformly arifes from the un-
found vegetables or putrid water, and never from clothes as the vehicles
of contagion, or bodies which may be difeafed on-board fuch veſfels;
our chief attention, inſtead of being confined to the perfons, ſhould be
directed to the ſtate of the cargoes; which by the bye ought always,
during the fickly months, to be examined before they are fuffered to
come fo near a city as that the noxious air can injure the inhabitants.

I have not annexed this appendix folely for the purpofe of adjuſting
the ettiquette of fcience, but in order to preferve the excellent preven-
tive obfervations contained in the reply to the requeſt of the governor
of Pennfylvania.

APPENDIX. 65

In this deciding upon the nature and origin of our late fever, we expect to administer consolation to our fellow citizens upon the cause of our late calamity, for in pointing out its origin to the senses, we are enabled immediately and certainly to prevent it. But while the only source of it is believed to be from abroad and while its entrance into our city is believed to be in ways so numerous and insidious as to elude the utmost possible vigilance of health officers, we are led in despair to consider the disease as removed beyond the prevention of human power or wisdom. It has been by adopting measures similar to those we have delivered for preventing pestilential diseases, that most of the cities of Europe, which are situated in warm latitudes, have become healthy in warm seasons and amidst the closest commercial intercourse with nations and islands constantly afflicted with those diseases. The extraordinary cleanliness of the Hollanders was originally imposed upon them by the frequency of pestilential fevers in their cities. This habit of cleanliness has continued to characterize those people after the causes which produced it have probably ceased to be known.

In thus urging a regard to the domestic sources of the yellow fever, we are actuated by motives of magnitude far beyond those which determine ordinary questions in science. Though we feel the strongest conviction that the value of property, the increase of commerce and the general prosperity of our city, will be eminently forwarded by the adoption of the foregoing propositions, yet these are but little objects in our view when compared with the prevention of the immense mass of distress which never fails to accompany a mortal epidemic. We consider ourselves moreover as deciding upon a question which is to affect the lives and happiness not only of the present inhabitants of Philadelphia, but of millions yet unborn in every part of the globe.

We are with the greatest respect,
Sir,
Your very humble servants,
Benjamin Rush,
Charles Caldwell,
William Dewees,
John Redman Coxe,
Phillip Syng Physick,
James Reynolds,
Francis Bowes Sayre,
John C. Otto,
William Boys,
Samuel Cooper,
James Stuart,
Felix Pascalis,
Joseph Strong.

His Excellency }
THOMAS MIFFLIN. }

E R R A T A.

IN page 7, line 35, for lotteals read latteals.
Page 15, line 11, for tread read trod. Line 20, for the the the Town, read the Town.
Page 16, line 11, for secendly read secondly.
Page 17, line 26, for antogonist read antagonist—line 33, for circumstences read circumstances.
Page 22, line 15, for whence read hence.
Page 26, line 23, for teneble read tenable.
Page 36, line 13, for strictun read strictum.
Page 38, line 10, for qud read quæ.
Page 42, line 17, for bill read bile—line 22, for is reason read is the reason, and for practicioners read practitioners.
Page 43, line 14, for rouvient read rouvrent.
Page 45, line 34, for ariæ read areæ.
Page 46, line 6, for eye-bales read eye-balls.
Page 51, line 4, for unweldiness read unweildiness—in line 5, for acid read acrid.
Page 52, line 20, for day read days.
Page 64, line 47, for ettiquette read etiquette.

Aliis, si qua sint, ignoscat benevolus lector.

www.ingramcontent.com/pod-product-compliance
Lightning Source LLC
Chambersburg PA
CBHW022007190326

41519CB00010B/1427